My World of Science

SOFT AND HARD

Angela Royston

Heinemann
LIBRARY

www.heinemann.co.uk/library
Visit our website to find out more information about **Heinemann Library** books.

To order:
☎ Phone 44 (0) 1865 888066
🗎 Send a fax to 44 (0) 1865 314091
💻 Visit the Heinemann Bookshop at www.heinemann.co.uk/library to browse our catalogue and order online.

First published in Great Britain by Heinemann Library, Halley Court, Jordan Hill, Oxford OX2 8EJ, part of Harcourt Education.

Heinemann is a registered trademark of Harcourt Education Ltd.

Editorial: Andrew Farrow and Dan Nunn
Design: Jo Hinton-Malivoire and
    Tinstar Design Limited (www.tinstar.co.uk)
Picture Research: Maria Joannou and Sally Smith
Production: Viv Hichens

Originated by Blenheim Colour Ltd
Printed and bound in China by
    South China Printing Company

ISBN 0 431 13740 4 (hardback)
07 06 05 04 03
10 9 8 7 6 5 4 3 2 1

ISBN 0 431 13746 3 (paperback)
08 07 06 05 04
10 9 8 7 6 5 4 3 2 1

British Library Cataloguing in Publication Data
Royston, Angela
Soft and hard. – (My world of science)
1. Hardness – Juvenile literature
I. Title
620.1'126

A full catalogue record for this book is available from the British Library.

Acknowledgements
The publishers would like to thank the following for permission to reproduce photographs:
Argos p. **18**; Chris Coggins p. **11**; Chris Honeywell p. **5**; Collections/Julie Hamilton p. **16**; Network Photographers p. **26**; Peter Gould p. **25**; Photodisc p. **28**; Pictor pp. **23**, **29**; Rupert Horrox p. **4**; Science Photo Library/Jerome Yeates p. **20**; Trevor Clifford pp. **6**, **7**, **8**, **9**, **10**, **12**, **13**, **14**, **15**, **17**, **19**, **21**, **22**, **27**; Trip/J. Greenberg p. **24**.

Cover photograph reproduced with permission of Trevor Clifford.

Every effort has been made to contact copyright holders of any material reproduced in this book. Any omissions will be rectified in subsequent printings if notice is given to the publishers.

Contents

Soft and hard 4

Which is softer? 6

Which is harder? 8

Choosing materials 10

Soft materials 12

Clothes 14

Filled with air 16

Hard materials 18

Plastic 20

Concrete and stone 22

Hard stones 24

Changing from soft to hard 26

Changing from hard to soft 28

Glossary 30

Answers 31

Index 32

Any words appearing in the text in bold, **like this**,
are explained in the Glossary.

Soft and hard

These teddy bears are soft. The **material** gives way under your fingers. If you **squeeze** them, your fingers will make a **dent**.

This toy train is hard. It will not change shape if you squeeze it. Your fingers will not leave a mark. Hard things make a sound when you tap them.

Which is softer?

Some things are softer than others.
The sponge cake is softer than the
playdough. But the cushion is the
softest thing in the picture.

This girl is testing these things to see which is the softest. She presses each one with her fingers. The softest one is the easiest one to **squeeze**.

Which is harder?

Wood is harder than an orange.
When the boy taps the wooden
block it makes a loud noise. But
when he taps the orange it doesn't.

All of these things feel hard to touch. The plastic plate is harder than the table mat. But which object is the hardest of them all? (Answer on page 31.)

Choosing materials

Some things are made of both hard and soft **materials**. The material used for the eyes of this toy **reindeer** is hard. But its body is made of soft material.

These chairs are made of hard wood. Wood makes the chairs strong. But each chair has a soft seat. These are more comfortable to sit on than wood.

Soft materials

Soft things can feel gentle against your skin. This **duvet** is filled with soft **material** to make it warm and **cosy**. The pillow and mattress are soft too.

Soft toys can feel good to cuddle.
This cat is filled with soft material and
covered with soft **fabric**. Which parts of
the cat are hard? (Answer on page 31.)

Clothes

It is more comfortable to wear something soft next to your skin than something hard and scratchy. This boy's soft, **fleecy** top is also light and warm.

This cap has a hard **shade** at the front. The shade protects the girl's face from the sun. The rest of the cap is softer. It fits around the girl's head.

Filled with air

This bouncy castle is made of hard, strong rubber, but it is filled with air. The air makes it soft to jump and fall about on.

Bicycle tyres are filled with air too. The tyres squash when the bicycle goes over a bump. This makes the bicycle more comfortable to ride.

Hard materials

Floors, walls and many pieces of furniture are made of hard **materials**. If they were not hard they would not keep their shape.

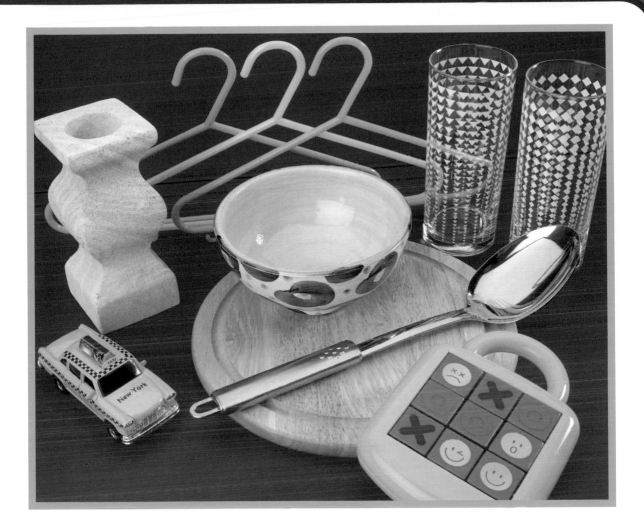

Many different materials are hard.
These things are made of stone, metal,
wood, glass, china and plastic. They
each have to be hard to do their job.

Plastic

A computer is made mostly of plastic. The case is hard so that it keeps its shape. The screen is covered with clear plastic.

Most kinds of plastic are hard. Hard things often last longer than soft things. One of these things is soft. Which is it? (Answer on page 31.)

Concrete and stone

This building is made of **concrete**. Concrete is made from cement powder and stones mixed with water. It becomes hard and strong when it is dry.

Stones and rocks are very hard.
Most stones and rocks have to be
dug from the ground with machines.
Stone is used to build homes.

Hard stones

This artist is working on a marble **statue**. Marble is one of the hardest stones. It is so hard, it can last for thousands of years.

Diamond is the hardest stone of all. It is so hard it can be cut only by another diamond. Tiny diamonds are used in dentists' **drills**.

Changing from soft to hard

Some things are sometimes soft and sometimes hard. When you ice a cake, the icing is soft enough to spread. A few hours later, it has set hard!

Most things become harder when they are frozen. Peas are usually soft. They become hard when they are stored in a freezer.

hard

soft

Changing from hard to soft

Some things become softer when they are heated. The **wax** in the top of the candle becomes soft and melts. The wax becomes hard again when it cools.

Fruit becomes softer as it ripens. The green berries on this blackberry bush are harder than the red berries. Only the black ones are soft and ripe.

Glossary

concrete material used to make buildings and paths
cosy snug and warm
dent small dip made by pressing something
drill machine used to make holes
duvet soft quilt that you sleep under in bed
fabric cloth
fleecy warm, light and fluffy
material the stuff that something is made of
reindeer type of deer with very large antlers
shade something that blocks the light
squeeze press between your hands or fingers
statue copy of a person, animal or other shape carved out of wood or stone. Some statues are made of metal.
wax material that is soft when heated. Most wax is made from oil, but some wax is made by bees.

Answers

page 9

The metal spoon is the hardest object of them all.

page 13

The eyes and nose of the cat are hard.

page 21

The sponge is soft.

Index

air 16–17
clothes 14–15
concrete 22
diamond 25
fabric 13
freezing 27
icing 26
marble 24
melting 28
metal 9
plastic 9, 20–21
ripe fruit 29
stone 23
wax 28
wood 8, 11